YOUR KNOWLEDGE HAS VALUE

- We will publish your bachelor's and master's thesis, essays and papers

- Your own eBook and book - sold worldwide in all relevant shops

- Earn money with each sale

Upload your text at www.GRIN.com and publish for free

Preparation and setting elements of hardened concrete

R.C. Nivita

Bibliographic information published by the German National Library:

The German National Library lists this publication in the National Bibliography; detailed bibliographic data are available on the Internet at http://dnb.dnb.de.

ISBN: 9783346829603
This book is also available as an ebook.

Print and binding: Books on Demand GmbH, Norderstedt, Germany
Printed on acid-free paper from responsible sources.

The present work has been carefully prepared. Nevertheless, authors and publishers do not incur liability for the correctness of information, notes, links and advice as well as any printing errors.

GRIN web shop: https://www.grin.com/document/1329685

HARDENED CONCRETE: BASICS OF PREPARATION AND SETTING

RC NIVITA

GCT, KOSI, BIHAR, INDIA

Abstract: This paper discusses preparation and setting elements of hardened concrete. The strength of hardened concrete is its resistance to forces. When a force compresses instead of stretches or pulls, this is known as compressive strength. The compressive strength of hardened concrete is regarded as its most critical quality indicator. Strength can indirectly reflect the majority of concrete properties directly related to the structure of hardened cement paste. Longer-lasting concrete is dense, compact, impermeable, weatherproof, and chemically resistant. However, employing more cement could make concrete more resistant to drying shrinkage and cracking, but also more durable. Depending on the material's workability, the type of compaction used to thoroughly compress the material will vary. In other words, the workability of concrete is contingent upon compaction and technique. Moreover, it relies on the working environment. These external vibrators are suitable with precast concrete. This approach delivers uniform vibration and reliably compacts prefabricated concrete. Prior to installation, concrete units are compacted using a vibrating table. Thus, the curing of concrete is completed by employing these approaches.

INTRODUCTION

Strength is one of the most fundamental qualities of hardened concrete, which reflects the material's resistance to the forces that are applied to it. Compressive strength is a type of strength that applies when the action of the force being applied is to compress rather than stretch or pull [1]. It is generally agreed upon that the most important characteristic of hardened concrete is its compressive strength, and this property is usually utilised as an indicator of the concrete's overall quality. Strength can provide an indirect indicator of the majority of other properties of concrete that are directly related to the structure of hardened cement paste [2]. These attributes include:

durability, workability, aesthetics, and durability. A more durable concrete is one that is dense, compact, impermeable, resistant to the effects of weathering, and resistant to the effects of certain chemicals [3]. The use of a larger cement content, on the other hand, can result in concrete that is stronger but also exhibits a greater tendency towards drying shrinkage and cracking [4].

Compressive strength is often coupled to a number of other desirable properties, including shear and tensile strengths, modulus of elasticity, bond strength, impact resistance, and so on [5]. Because compressive strength can be easily evaluated on specimens of standard sizes, it can be stated as a criterion for analysing the effect that any variable has on the quality of concrete [6]. However, the qualities of concrete are tested in a variety of scenarios, each of which produces a unique set of results. As a consequence of this, the method of testing, the size of the specimen, the rate of loading, and other factors are defined when testing concrete in order to reduce the amount of variation in the results of the tests [7]. When attempting to measure the value of a single property of hardened concrete, statistic-based approaches are typically the methods of choice [8].

THE STRENGTH
Concrete's compressive strength can be described as the force per unit area of cross-section that causes a specimen to fail under uniaxial compression at a particular rate of loading [9]. This is referred to as the "failure load." The amount of force that concrete can exert is measured in N/mm2 [10]. After casting, the compressive strength of the concrete is measured after 28 days to be used as a criterion for identifying the quality of the concrete. The term for this type of grade is concrete grade. The use of cubes measuring 150 millimetres squared is required by I.S. 456 – 2000. [11]

STRETCHING STRENGTH

Between 8 and 12 percent of the material's compressive strength is comprised of the tensile strength of the concrete. In most cases, a median number that is equal to 10 percent is applied [12].

SHEAR STRENGTH

In concrete, tensile and compressive stresses are balanced out by bending and shear forces. The shear failures that result from the resulting diagonal tension [13]. The shear strength is typically between 12 and 13 percent of the material's compressive strength [14].

BOND STRENGTH

The resistance of concrete to the sliding of reinforcing bars that are implanted in the material is referred to as bond strength [15]. The strength of the bond is due to the adhesion of the solidified cement paste as well as the friction that occurs between the concrete and the steel. In addition to this, it is impacted by the proportional contraction of concrete and steel [16]. The typical bond strength is equal to ten percent of the compressive strength of the material.

In order to properly hydrate one bag of cement, you will need between 25 and 28 litres of water [17].

FACTS REGARDING CEMENT AND CONCRETE

1. The colour of concrete does not have any influence on the quality of the concrete.

2. Mortar or concrete should be used as soon as it is practically possible after water has been added to the mixture. As soon as the cement is exposed to water, the process of hydration can begin. As the hydration process proceeds, the cement paste will gradually become more rigid and less fluid [18]. After this, the concrete shouldn't be moved or messed with in any way. This takes somewhere between 45 and 50 minutes, on average [19].

3. The pressure unit mega Pascal is abbreviated as MPa, which is a shorter version of the full abbreviation. The pressure is equal to 10 kg/cm2 for every 1 MPa. The amount of pressure that a standard cube is able to resist is used as a measure of the strength of concrete and cement [20]. Ordinary Portland Cement, which is sometimes referred to as OPC in some circles, can be purchased in one of three grades: 33, 43, or 53. After 28 days of ordinary curing circumstances, a mortar cube made with conventional cement and sand would have

a minimum strength of 43 MPa, which is equivalent to 430 kg/cm2, if it are made with 43

grade cement [21].

Concrete's compressive strength is determined by a number of factors, including the following:

(i)   weight-to-volume ratio

(ii)   Components of cement

(iii)   The properties of the aggregates

(iv)   Interval during which we mix

(v)   The degree of the compactness

(vi)   Temperature and cure period

(vii)   How old is the cement?

(viii)   Air entertainment

(ix)   The parameters of the investigation

PRECAUTIONS FOR USING WATER IN CONCRETE

• Using water that is fit for human consumption is advantageous. • The water in question must be

free of pollutants and other potentially harmful chemicals [22].

It is not recommended to drink seawater.

• Water that is suitable for mixing can also be used for the curing process.

• It is recommended to use the least amount of mixing water that is compatible with the degree of

workability that is required to assist easy placing and compaction of concrete. • Make ensure that

water is metered and added [23].

• In order for a structure to maintain its strength over time, it must have a water-to-cement ratio

that is relatively low [24].

COMMON REASONS FOR UNSATISFIED WORK IN CONCRETE

• Using an excessive amount of water for mixing, using an inadequate amount of water for mixing, or adding water recklessly during mixing

• An inadequate mixing of the aggregate with the cement [25].

• An incorrect grading of the aggregates, which might cause bleeding or segregation in the concrete [26].

• The concrete isn't packed down tightly enough. • Using concrete that had already started to set.

• Placing concrete on a dry basis without first thoroughly wetting the base with water before doing so [27].

• The utilisation of aggregate that has been contaminated with soil or water that contains earth, clay, or lime.

• An excessive amount of troweling is done on the surface of the concrete [28].

During the first ten days after placement of the concrete, leaving the surface exposed to the sun and wind without protecting it or maintaining its moisture through the use of proper curing processes [29].

Joints in construction are the joints that are given between successive pours of concrete that have been carried out after a certain amount of time has passed. It is imperative that every effort be made to reduce the number of construction joints and, if at all feasible, do away with them entirely [30]. Because the presence of these joints creates a plane of weakness inside the concrete body, the positioning of those joints needs to be planned in advance in order to ensure that they are subjected to the least amount of bending moment and shear force that is feasible [31].

Concrete is used in every part of the construction process, including the columns, beams, and slabs (M20). It is fully compressed by using portable mechanical vibrators, which allowed for more efficiency [32]. It is imperative that the concrete not be over-vibrated to the point that it became separated into its constituent parts. The arrangement of the layers of concrete is such that the bottom layer does not completely harden prior to the insertion of the top layer in the structure [33]. The vibrators maintain a sufficient state of agitation throughout the entirety of the mass of concrete that is undergoing treatment. This allows for de-aeration and effective compaction to be accomplished at a level that is commensurate with the quantity of concrete that is delivered by the mixers [34]. The vibrators are set up in such a way that, at the moment when the mass is being placed, the centre of the mass that is being compacted is in close proximity to the centre of the device that is vibrating. It is important to avoid shaking the reinforcement in order to prevent it from becoming loose. Prior to the beginning of first setting, or within thirty minutes of adding water to the dry mixture, compaction needs to be done so that the mixture may be used [35].

In order to stop the constituent parts from becoming separated, the concrete is poured into its final position [36]. The shuttering for the columns and walls is managed in such a way that the vertical drop of concrete does not exceed 1.5 metres at any one time. In the instance of the slab and beam concreting, the pipe is brought straight from the facility that did the batching to the area that is the closest to it [37].

COMPACTION

There are three steps to the creation of green concrete: solids, liquids, and air. In order to make concrete impenetrable and attain its greatest strength, it is vital to remove any air pockets from the mass of concrete while it is still in a flexible state [38]. This will allow the concrete to fulfil its full potential. The strength of the concrete suffers dramatically if all of the air is not entirely removed from the mixture. It has been demonstrated that the amount of strength lost due to the presence of 5% cavities is around 30%, while the amount of strength lost due to the presence of 10% voids is nearly 50% [39]. The reduction of air bubbles and the distribution of a sufficient amount of fine material on the surface and against the forms are both accomplished through the process of compaction. It is possible to use hand tools such as steel rods, paddling sticks, and tampers; nevertheless, mechanical vibrators are the instrument of choice because of their greater efficiency [40]. Any device that compacts the material must be able to reach the base of the form while also being small enough to fit in the spaces between the reinforcing bars. Because the durability of the concrete component is dependent on the reinforcing steel being correctly positioned, extreme caution is required to ensure that it is not moved in any way [41].

When working with reinforced concrete, it is essential to employ iron rods in the compacting process. If the total thickness of the concrete layers is greater than 15 centimetres, the most efficient method for effectively compacting the concrete is to consolidate each layer individually before installing the subsequent layer [42]. This will ensure that the top surface of each layer is level and relatively smooth before the installation of the subsequent layer. To guarantee that the layers properly adhere to one another, attention should be taken during the tamping process to ensure that the rod penetrates not only the complete layer of the layer that is laid immediately before it but also a piece of the layer directly below it [43]. Second, the reinforcing material and the formwork should remain in their respective locations and should not be moved.

TECHNIQUES

The workability of the material will define the method of compaction that will be used to ensure that the material is completely compressed. To put it another way, the needed workability of the concrete is dependant upon the quantity of compaction as well as the technique used [44]. In addition to that, it will be dependent on the working conditions. Typical procedures for compacting materials include the following:

(1). Hand compaction can be accomplished by a variety of techniques, including tamping, rodding, and pounding [45].

In most cases, the process of tamping is utilised to compact concrete for slabs and other surfaces that are comparable; roding is utilised for the compacting of thin vertical sections; and hammering is utilised for the compacting of massive plain concrete buildings [46]. This method is utilised for members that have reinforcement, such as the pavement, narrow and deep members, and other similar members. The use of rammers and iron rods is necessary to achieve this goal [47]. Layers of mass concrete shouldn't be any thinner than 30 centimetres, and they should be compressed using templates or light rammers. For the purpose of compacting reinforced concrete, rods of iron are utilised. If the thickness of the concrete layers is to be greater than 15 centimetres. The most efficient method for correctly compacting concrete is to consolidate each layer individually before placing the subsequent layer [8]. This should be done in such a way that the top surface of each layer becomes level and relatively smooth before the placement of the subsequent layer. When you are tamping, you need to make sure that the tamping rod goes all the way through the layer that is just laid and continues into the one underneath it. This will ensure that there is a strong connection between the layers. Second, the reinforcing material and the formwork should remain in their respective locations and should not be moved [29].

(2). Vibrators are used to bring about the process of mechanical compaction, also known as "mechanical compression." It is well accepted that the use of vibration to compact concrete is a necessary step in any substantial operation. This is especially true in circumstances in which the reinforcements are crowded or the member is required to have an exposed concrete surface finish. Concrete mixtures that are too rigid to be compacted by hand can be easily compacted using a machine. If the concrete is compacted using vibrations, the rapid vibrations that are sent from the vibrator to the particles in the concrete make the concrete more fluid. The particles settle into a position that is more stable as a result of the vibrations, the concrete fills all of the available space, and the present is forced to the surface, which results in concrete that is dense and long-lasting [30].

TYPES OF VIBRATORS

The following types of vibrations are typically utilised during the process of compacting concrete:

1. Internal vibrators

2. External vibrators

3. Surface vibrators

4. Vibrating table

An internal vibrator has a vibrating head that looks like a metal road and is sunk to the full depth of the concrete layer. It is a sort of vibrator that makes direct contact with the concrete and is hence regarded as the most efficient type of vibrator. It is also known as a poker or needle vibrator [41]. External vibrators are pressed up against the concrete formwork, and vibrations that cause concrete to compact are communicated to the material via the formwork. These are also known as form vibrators in some circles. Because the vibrator is firmly attached to the formwork, which is lying on an elastic spot, the vibrations are transmitted to both the form and the concrete [12]. In the event that vibration consumes a large portion of the work that is being done, the efficiency of the system will be reduced. Platform-based surface vibrators are often used to compact and complete a variety of construction projects, including highways, bridges, and more. These are the types of outside vibrators that are suitable for applications involving precast concrete. In addition to being a dependable way for compacting precast concrete, this technique also offers the benefit of providing uniform vibration. A vibrating table is used to compact the precast units before they are installed. Surface vibrators are utilised in locations such as dams and walls that are quite thick when there is a huge horizontal surface area present [3]. There are many other kinds of surface vibrators, but the one that looks like a pen is by far the most prevalent form. When concrete is placed on top of these tables, a process known as mechanical compaction takes place, which has a great number of benefits. Because every kind of vibrator comes with its own set of advantages and disadvantages, it is essential to make an educated decision whenever you have to pick one over the other. When compacting concrete by vibration, careful planning is essential to ensure success. If the thickness of such concrete is greater than 5 centimetres, then concrete segregation will occur, which is something that must not take place under any circumstances [24].

After the concrete surface is adequately settled, all of the exposed surfaces of the RCC lintels, beams, columns, and other structural elements are plastered to match the plastered face of the neighbouring walls [25]. A device that mixes multiple components in order to manufacture concrete is known as a concrete plant and is also referred to as a batch plant on occasion. Sand, water, aggregate (rocks, gravel, etc.), fly ash, potash, cement, and any other elements that are used in the construction of concrete are examples of some of these inputs. There are two varieties of concrete plants: ready-mix plants and central mix plants [26]. Both may produce concrete. Mixers (either tilt-up or horizontal (or in some cases, both)), cement batchers, aggregate batchers, conveyors, radial stackers, aggregate bins, cement bins, heaters, chillers, cement silos, batch plant controls, and dust collectors can all be found in a concrete plant. These parts and accessories are just some of the many that can be found in a concrete plant (to minimise environmental pollution) [45].

CONCLUSION

This paper deals with aspects related to preparation and setting of hardened concrete. Hardened concrete's strength is its resistance to forces. Compressive strength occurs when a force compresses rather than stretches or pulls. Hardened concrete's compressive strength is considered its most essential quality indicator. Strength can indirectly indicate most concrete qualities directly related to hardened cement paste structure. Dense, compact, impermeable, weatherproof, and chemical-resistant concrete lasts longer. However, using more cement might make concrete stronger but more prone to drying shrinkage and cracking. The method of compaction utilised to fully compress the material depends on its workability. In other words, concrete workability depends on compaction and technique. It also depends on working conditions. These are precast concrete-compatible exterior vibrators. This method provides homogeneous vibration and compacts precast concrete reliably. A vibrating table compacts concrete units before installation. Thus, adopting these techniques completes the setting of concrete.

References

[1]   R. Garg, R. Garg, and N. O. Eddy, "Microbial induced calcite precipitation for self-healing of concrete: a review," *J. Sustain. Cem. Mater.*, 2022, doi: 10.1080/21650373.2022.2054477.

[2]   H. Liu, Q. Li, D. Su, G. Yue, and L. Wang, "Study on the influence of nanosilica sol on the hydration process of different kinds of cement and mortar properties," *Materials (Basel).*, vol. 14, no. 13, 2021, doi: 10.3390/ma14133653.

[3]   S. Mandal, J. K. Singh, D. E. Lee, and T. Park, "Effect of phosphate-based inhibitor on corrosion kinetics and mechanism for formation of passive film onto the steel rebar in chloride-containing pore solution," *Materials (Basel).*, vol. 13, no. 16, pp. 7–9, 2020, doi: 10.3390/MA13163642.

[4]   E. Najaf, M. Orouji, and S. M. Zahrai, "Improving nonlinear behavior and tensile and compressive strengths of sustainable lightweight concrete using waste glass powder, nanosilica, and recycled polypropylene fiber," *Nonlinear Eng.*, vol. 11, no. 1, pp. 58–70, 2022, doi: 10.1515/nleng-2022-0008.

[5]   Antoni, J. G. Halim, O. C. Kusuma, and D. Hardjito, "Optimizing Polycarboxylate Based Superplasticizer Dosage with Different Cement Type," *Procedia Eng.*, vol. 171, pp. 752–759, 2017, doi: 10.1016/j.proeng.2017.01.442.

[6]   H. Sharma, R. Garg, D. Sharma, M. umar Beg, and R. Sharma, "Investigation on Mechanical Properties of Concrete Using Microsilica and Optimised dose of Nanosilica as a Partial Replacement of Cement," *Int. J. Recent Reasearch Asp.*, vol. 3, no. 4, pp. 23–29, 2016.

[7]   R. Garg, R. Garg, B. Chaudhary, and S. Mohd. Arif, "Strength and microstructural analysis of nano-silica based cement composites in presence of silica fume," *Mater. Today Proc.*, vol. 46, pp. 6753–6756, 2020, doi: 10.1016/j.matpr.2021.04.291.

[8]   H. B. Tran, V. B. Le, and V. T. A. Phan, "Mechanical properties of high strength concrete

containing nano sio2 made from rice husk ash in Southern Vietnam," *Crystals*, vol. 11, no. 8, 2021, doi: 10.3390/cryst11080932.

[9] Rishav Garg, Manjeet Bansal, and Yogesh Aggarwal, "Split Tensile Strength of Cement Mortar Incorporating Micro and Nano Silica at Early Ages," *Int. J. Eng. Res.*, vol. V5, no. 04, pp. 16–19, 2016, doi: 10.17577/ijertv5is040078.

[10] Atta-ur-Rehman, A. Qudoos, S. H. Jakhrani, H. G. Kim, and J.-S. Ryou, "Influence of Nano-silica on the Leaching Attack upon Photocatalytic Cement Mortars," *Int. J. Concr. Struct. Mater.*, vol. 13, no. 1, p. 35, Dec. 2019, doi: 10.1186/s40069-019-0348-x.

[11] D. Prasad Bhatta, S. Singla, and R. Garg, "Experimental investigation on the effect of Nano-silica on the silica fume-based cement composites," *Mater. Today Proc.*, vol. 57, pp. 2338–2343, 2022, doi: 10.1016/j.matpr.2022.01.190.

[12] P. Smarzewski, "Influence of silica fume on mechanical and fracture properties of high performance concrete," *Procedia Struct. Integr.*, vol. 17, pp. 5–12, 2019, doi: 10.1016/j.prostr.2019.08.002.

[13] A. Singh, S. Singla, R. Garg, and R. Garg, "Performance analysis of Papercrete in presence of Rice husk ash and Fly ash," *IOP Conf. Ser. Mater. Sci. Eng.*, vol. 961, p. 012010, Nov. 2020, doi: 10.1088/1757-899X/961/1/012010.

[14] H. T. Le and H. M. Ludwig, "Effect of rice husk ash and other mineral admixtures on properties of self-compacting high performance concrete," *Mater. Des.*, vol. 89, pp. 156–166, 2016, doi: 10.1016/j.matdes.2015.09.120.

[15] J. J. Chen, P. L. Ng, L. G. Li, and A. K. H. Kwan, "Production of High-performance Concrete by Addition of Fly Ash Microsphere and Condensed Silica Fume," *Procedia Eng.*, vol. 172, pp. 165–171, 2017, doi: 10.1016/j.proeng.2017.02.045.

[16]   M. Benaicha, A. Hafidi Alaoui, O. Jalbaud, and Y. Burtschell, "Dosage effect of superplasticizer on self-compacting concrete: Correlation between rheology and strength," *J. Mater. Res. Technol.*, vol. 8, no. 2, pp. 2063–2069, 2019, doi: 10.1016/j.jmrt.2019.01.015.

[17]   G. M. Fani, S. Singla, R. Garg, and R. Garg, "Investigation on Mechanical Strength of Cellular Concrete in Presence of Silica Fume," *IOP Conf. Ser. Mater. Sci. Eng.*, vol. 961, no. 1, p. 012008, Nov. 2020, doi: 10.1088/1757-899X/961/1/012008.

[18]   C. Xupeng, S. Zhuowen, and P. Jianyong, "Study on Metakaolin Impact on Concrete Performance of Resisting Complex Ions Corrosion," *Front. Mater.*, vol. 8, no. December, pp. 1–12, 2021, doi: 10.3389/fmats.2021.788079.

[19]   R. Garg, T. Biswas, M. D. Alam, A. Kumar, A. Siddharth, and D. R. Singh, "Stabilization of expansive soil by using industrial waste," *J. Phys. Conf. Ser.*, vol. 2070, no. 1, p. 012238, Nov. 2021, doi: 10.1088/1742-6596/2070/1/012238.

[20]   Y. C. Ersan, "Overlooked Strategies in Exploitation of Microorganisms in the Field of Building Materials," Springer Singapore, 2019, pp. 19–45. doi: 10.1007/978-981-13-0149-0_2.

[21]   L. Senff, D. Hotza, and J. a Labrincha, "Effect of diatomite addition on fresh and hardened properties of mortars investigated through mixture experiments," *Adv. Appl. Ceram.*, vol. 110, no. 3, pp. 142–150, Apr. 2011, doi: 10.1179/1743676110Y.0000000009.

[22]   R. Garg, R. Garg, and N. O. Eddy, "Influence of pozzolans on properties of cementitious materials: A review," *Adv. Nano Res.*, vol. 11, no. 4, pp. 423–436, 2021, doi: 10.12989/anr.2021.11.4.423.

[23]   A. Nazari, S. Riahi, S. Riahi, S. F. Shamekhi, and A. Khademno, "Mechanical properties of cement mortar with Al2O3 nanoparticles," *J. Am. Sci.*, vol. 6, no. 4, pp. 94–97, 2010.

[24]   K. Kumar, M. Bansal, R. Garg, and R. Garg, "Mechanical strength analysis of fly-ash based

concrete in presence of red mud," *Mater. Today Proc.*, vol. 52, no. xxxx, pp. 472–476, 2022, doi: 10.1016/j.matpr.2021.09.233.

[25] U. Ubinayaa, D. Chetha, S. Chathuska, N. Praneeth, R. Vimantha, and K. K. Wijesundara, "Improving the properties of concrete using carbon nanotubes," *SAITM Res. Symp. Eng. Adv.*, vol. 2014, p. 4, 2014.

[26] T. Parhizkar, A. M. R. Ghasemi, and A. A. Ramezanianpour, "Properties of the New Type of HPC in Simulated Conditions of Persian Gulf".

[27] R. Garg *et al.*, "Mechanical strength and durability analysis of mortars prepared with fly ash and nano-metakaolin," *Case Stud. Constr. Mater.*, vol. 18, no. December 2022, p. e01796, 2023, doi: 10.1016/j.cscm.2022.e01796.

[28] R. P. S. Kushwah and O. Prakash, "Utilization of 'marble slurry' in cement mortar," *Int. J. Eng. Sci. Res. Technol.*, vol. 5, no. 9, pp. 640–645, 2016, doi: 10.5281/zenodo.155087.

[29] J. Yu, M. Zhang, G. Li, J. Meng, and C. K. Y. Leung, "Using nano-silica to improve mechanical and fracture properties of fiber-reinforced high-volume fly ash cement mortar," *Constr. Build. Mater.*, vol. 239, p. 117853, Apr. 2020, doi: 10.1016/j.conbuildmat.2019.117853.

[30] A. Singh, A. Garg, and R. Garg, "Influence of Nano-Silica and Ground Granulated Blast Furnace Slag on Cement Using Statistical," 2018.

[31] D. G. Leo Samuel, K. Dharmasastha, S. M. Shiva Nagendra, and M. P. Maiya, "Thermal comfort in traditional buildings composed of local and modern construction materials," *Int. J. Sustain. Built Environ.*, vol. 6, no. 2, pp. 463–475, 2017, doi: 10.1016/j.ijsbe.2017.08.001.

[32] B. B. Das and A. Mitra, "Nanomaterials for Construction Engineering-A Review," *Int. J. Mater. Mech. Manuf.*, vol. 2, no. 1, pp. 41–46, 2014, doi: 10.7763/ijmmm.2014.v2.96.

[33] Y. Qing, Z. Zenan, K. Deyu, and C. Rongshen, "Influence of nano-SiO2 addition on

properties of hardened cement paste as compared with silica fume," *Constr. Build. Mater.*, vol. 21, no. 3, pp. 539–545, Mar. 2007, doi: 10.1016/j.conbuildmat.2005.09.001.

[34]   M. Seifan, A. Ebrahiminezhad, Y. Ghasemi, A. K. Samani, and A. Berenjian, "The role of magnetic iron oxide nanoparticles in the bacterially induced calcium carbonate precipitation," *Appl. Microbiol. Biotechnol.*, vol. 102, no. 8, pp. 3595–3606, 2018, doi: 10.1007/s00253-018-8860-5.

[35]   Y. Kocak, "A study on the effect of fly ash and silica fume subsituted cement paste and mortars," *Sci. Res. Essays*, vol. 5, no. 9, pp. 990–998, 2010.

[36]   R. Sharma, D. Sharma, and R. Garg, "Development of Self Compacted High Strength Concrete using Steel Slag as Partial Replacement of Natural Fine Aggregate," pp. 96–99, 2016.

[37]   N. Stevulova, I. Schwarzova, V. Hospodarova, and J. Junak, "Implementation of waste cellulosic fibres into building materials," *Chem. Eng. Trans.*, vol. 50, pp. 367–372, 2016, doi: 10.3303/CET1650062.

[38]   A. Albidah, M. Alghannam, H. Abbas, T. Almusallam, and Y. Al-Salloum, "Characteristics of metakaolin-based geopolymer concrete for different mix design parameters," *J. Mater. Res. Technol.*, vol. 10, pp. 84–98, 2021, doi: 10.1016/j.jmrt.2020.11.104

[39]   A. Garg, A. Singh, and R. Garg, "Effect of Rice Husk Ash & Cement on CBR values of Clayey soil," *Int. J. Eng. Res. Appl.*, vol. 3, no. 1, pp. 1710–1717, 2013.

[40]   K. Kumar, M. Bansal, R. Garg, and R. Garg, "Penetration and strength analysis of pervious concrete," *J. Phys. Conf. Ser.*, vol. 2070, no. 1, 2021, doi: 10.1088/1742-6596/2070/1/012244.

[41]   M. Jalal, A. R. Pouladkhan, H. Norouzi, and G. Choubdar, "Chloride penetration, water absorption and electrical resistivity of high performance concrete containing nano silica and silica fume," *J. Am. Sci.*, vol. 8, no. 4, pp. 278–284, 2012.

[42]   J. Liu, H. Jin, C. Gu, and Y. Yang, "Effects of zinc oxide nanoparticles on early-age hydration and the mechanical properties of cement paste," *Constr. Build. Mater.*, vol. 217, pp. 352–362, 2019, doi: 10.1016/j.conbuildmat.2019.05.027.

[43]   R. Singh and S. Goel, "Experimental investigation on mechanical properties of binary and ternary blended pervious concrete," *Front. Struct. Civ. Eng.*, vol. 14, no. 1, pp. 229–240, 2020, doi: 10.1007/s11709-019-0597-4.

[44]   E. Molaei Raisi, J. Vaseghi Amiri, and M. R. Davoodi, "Mechanical performance of self-compacting concrete incorporating rice husk ash," *Constr. Build. Mater.*, vol. 177, pp. 148–157, 2018, doi: 10.1016/j.conbuildmat.2018.05.053.

[45]   H. S. Gökçe, D. Hatungimana, and K. Ramyar, "Effect of fly ash and silica fume on hardened properties of foam concrete," *Constr. Build. Mater.*, vol. 194, pp. 1–11, 2019, doi: 10.1016/j.conbuildmat.2018.11.036.